NATO

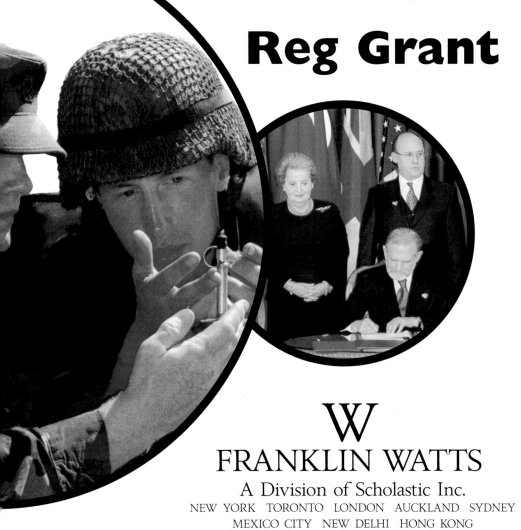

Reg Grant

W

FRANKLIN WATTS

A Division of Scholastic Inc.

NEW YORK TORONTO LONDON AUCKLAND SYDNEY

MEXICO CITY NEW DELHI HONG KONG

DANBURY, CONNECTICUT

Picture credits:
Cover: NATO br; Popperfoto l (Reuters/Kevin Capon), r (Reuters/Yannis Behrakis)
Inside: Corbis p. 4 b (Bettmann); Corbis-Bettmann p. 26 (AFP); Hulton Getty pp. 8 t (MPI/Truman Library), 14 cr, 14 br & bl (MPI Archives); Magnum Photos p. 12 b (Philip Jones Griffiths); NATO 5 t; Popperfoto pp. 11 (Reuters/Viktor Korotayev), 1 r (Reuters/Jeff Taylor), 2 t (Reuters/Paolo Cocco), 2/3 b (Reuters/Yannis Behrakis), 4 t (Reuters), 5 b (AFP/Persson), 7 (Gary Hershorn), 8 b, 9 t, 9 b, 10, 11 t (Reuters/Dswa-Dasiac), 11 b, 12 tl, 12/13, 13, 14 t, 15 t (AFP/Alexander Nemenov), 15 b (AFP/Gennady Tamarin), 16 (Reuters/ Paolo Cocco), 17 (Reuters/Viktor Korotayev), 18 t (Reuters/Danilo Krstanovic), 18 b (Reuters/Jeff Taylor), 19 (Reuters/Benoit Doppagne), 20 (EPA/S. Pikula), 21 t (Reuters/Benoit Doppagne), 21 b (Reuters/Yannis Behrakis), 22 t (Reuters), 22 b (Reuters/Yannis Behrakis), 23 t (Reuters/Paul McErlane), 23 b (Reuters/Petar Kujundzic), 24 t & b (Reuters/Emil Vas), 25 (Reuters/Chris Helgren), 27 (Reuters), 28 cr & c (Reuters), 29 t & b (Reuters); Topham Picturpoint pp. 3, 6 b.

First published in 2001 by Franklin Watts

Series Editor: Anderley Moore
Designer: Simon Borrough
Picture Research: Sue Mennell
Franklin Watts would like to thank Chris Bennett, the editor of NATO Review, for his help and advice in compiling this book.

First American edition 2001 by
Franklin Watts
A Division of Scholastic Inc.
90 Sherman Turnpike
Danbury, CT 06816

Visit Franklin Watts on the Internet at:
http://publishing.grolier.com

Catalog-in-Publication Data is available from the Library of Congress
ISBN   0-531-14622-7

Contents

1. What Is NATO?   **4**

2. How NATO Was Formed   **8**

3. NATO in the Cold War   **11**

4. From Confrontation to   **15**
   Cooperation

5. NATO in Kosovo   **20**

6. The Future of NATO   **25**

Glossary   **30**

Useful Information   **31**

Index   **32**

1. What Is NATO?

The North Atlantic Treaty Organization (NATO) was set up in 1949 by twelve countries in Europe and North America, including the United States and the United Kingdom. More European countries have joined over the years, and by 2000, NATO had nineteen members.

Joining Forces

NATO is a political and military alliance. It was formed soon after World War II, when the threat of further conflict seemed very real. The NATO countries wanted to defend themselves against a possible attack by the Soviet Union (the communist country that included present-day Russia and Ukraine).

◀ *Representatives of NATO countries and other friendly nations meet at NATO's headquarters in Brussels in 1995.*

▶ *NATO was formed after war had devastated Europe.*

 Checklist

NATO Members

By the year 2000, nineteen countries in Europe and North America were members of NATO:

Country	Year joined
Belgium	1949
Canada	1949
Czech Republic	1999
Denmark	1949
France	1949
Germany, West	1955
Greece	1952
Hungary	1999
Iceland	1949
Italy	1949
Luxembourg	1949
Netherlands	1949
Norway	1949
Poland	1999
Portugal	1949
Spain	1982
Turkey	1952
United Kingdom	1949
United States	1949

The NATO member countries agreed that an attack on any one of them would be seen as an attack on them all. They formed a joint military command, so that were one country attacked, all the other countries would launch a coordinated defense of that country. Any decision of this kind would be made jointly by the governments of the NATO member countries. Although NATO was officially an alliance of equal, independent countries, the United States has often taken the leadership role in the organization because of its huge size and strength.

With time, NATO became a very powerful military alliance. Until the early 1990s, NATO was in confrontation with the countries united under the Warsaw Pact, an alliance of East European countries led by the Soviet Union. This confrontation, which involved no actual fighting, was known as the Cold War.

▲ The NATO symbol has represented security for millions of people since 1949.

Promoting Peace and Democracy

NATO has a wider purpose than that of a defensive military alliance. Its declared goals are to defend democracy and freedom and promote peace and stability throughout Europe and North America. Since the end of communist rule in Eastern Europe in the late 1980s and the collapse of the Soviet Union in 1991, NATO countries have had no enemy that could seriously threaten them. As a result, defense has become less important, and the wider goals of promoting peace and democracy have come to the forefront of the organization's work.

In the 1990s, NATO led peacekeeping operations in war-torn Bosnia-Herzegovina, and intervened in a conflict in the southern Serbian province of Kosovo. NATO also set out to establish good relations with its former enemies, particularly with Russia and Ukraine.

▶ Civilians in the Bosnian capital of Sarajevo seek shelter from the fire of a Serbian sniper while a Bosnian soldier shoots back. NATO peacekeepers helped stop this kind of fighting in Bosnia.

Problem

In the Name of Democracy

One of NATO's major problems was that it claimed to defend freedom and democracy, but some member countries were not democracies — Greece from 1967 to 1974, and Turkey for much of the time. Since the mid-1970s, however, all NATO members have been stable democracies, and most have fairly good human rights records. After the death of General Franco in 1975, Spain became a democracy, and the Spanish people voted to join NATO in a referendum in 1982.

Problem

An Aggressive Presence?

NATO is not a popular organization with everyone. At different times many people, especially in Europe, have been hostile to it for a variety of reasons. They have, for example, seen it as a means for the United States to dominate Europe, or as an aggressive and bullying military power. But NATO sees itself as a force for good. Its supporters say it has maintained peace, preventing any major war from happening in Europe for over half a century.

The Structure of NATO

NATO's headquarters are in Brussels, Belgium. About three thousand people work there. Each member country has an ambassador at NATO headquarters, and these ambassadors meet at least once a week as the North Atlantic Council (NAC). The council is the main decision-making body of the organization. Occasionally, the foreign ministers of the member countries or even their heads of government meet in a special session of the council. Meetings of the council are chaired by NATO's secretary-general, who is selected from the member countries. In 1999, the United Kingdom's Lord Robertson became secretary-general of NATO, a post he will hold for four years.

▼ *Flags of member countries fly outside the NATO headquarters building in Brussels, Belgium. At the center is the organization's own flag.*

▲ In April 1999 in Washington, D.C., U.S. president Bill Clinton speaks at a ceremony celebrating the fiftieth anniversary of the founding of NATO.

From the outside, NATO's decision-making process seems to take forever. This is because it is based on consensus, meaning that all decisions have to be unanimous. That way, the independence of each member country is preserved, and every decision is backed by all NATO members.

The military side of NATO is headed by a military committee made up of chiefs of staff from each member country. NATO does not have independent armed forces. The troops and equipment available to NATO remain under the control of their own government until they are needed. Then, by agreement with each government, NATO forces begin the next task, whether it be defense, peacekeeping, or responding to a natural disaster.

NATO countries officially marked their commitment to each other's defense by signing the North Atlantic Treaty. It was signed in the capital of the United States, Washington, D.C., on April 4, 1949, and the North Atlantic Treaty Organization (NATO) came into being. To understand the need for such an organization, it is important to consider the events that led up to its formation.

▲ *U.S. president Harry S. Truman signs the North Atlantic Treaty in 1949, committing his country to the defense of Western Europe.*

▼ *Soviet troops in Berlin, 1948. After the war, Germany split in two: West was democratic, and East was communist. Germany's capital, Berlin, was also split.*

Europe's Iron Curtain

During World War II, which ended in 1945, an alliance that included the Soviet Union, the United Kingdom, and the United States defeated Nazi Germany. After the war, however, the Soviet Union fell out with its former allies in the West. During the war, the Soviet army had occupied Europe as far west as central Germany. After the war, the areas under Soviet occupation, including Czechoslovakia, Hungary, and Poland, were put under Soviet-style communist governments that had little respect for human rights. Europe was now divided in two: Soviet communism in the east and liberal democracy in the west — a divide that became known as the Iron Curtain.

Compared to the Soviet forces, the armies of West European countries such as France and the United Kingdom were weak. If the Soviet Union attacked them, they would not be able to defend themselves successfully. West European leaders, such as British foreign secretary Ernest Bevin, sought to enlist the help of another powerful liberal democracy, the United States. They hoped the United States would commit itself to defending Western Europe against a possible Soviet attack.

▲ Ernest Bevin (above), British foreign secretary (1945–51), played a large part in the creation of NATO.

◀ In 1948–49, the Soviets tried to take over Berlin. The communist blockade was broken by British and American aircraft flying in supplies to West Berlin.

The United States and NATO

At the end of World War II, the United States had withdrawn most of its forces from Europe. The United States traditionally avoided permanent alliances with foreign countries.

By 1947, however, the U.S. government, led by President Harry S. Truman, had become concerned about the possible spread of communism worldwide. The United States did not want to see Western Europe come under communist rule, either through a Soviet invasion or a communist revolution in Western Europe.

In June 1947, the U.S. Congress passed the Marshall Plan, under which the United States provided massive economic aid for the rebuilding of democratic Western Europe.

European countries made the first moves toward a military alliance with the United States. In March 1948, the Brussels Treaty created the Western Union, a defensive alliance between Belgium, France, Luxembourg, the Netherlands, and the United Kingdom. The Western Union then began negotiations to expand the alliance to include Canada and the United States, as well as other European countries. The founding members of the alliance that became NATO were Belgium, Canada, Denmark, France, Iceland, Italy, Luxembourg, the Netherlands, Norway, Portugal, the United Kingdom, and the United States.

▼ *The communist threat was embodied by the Soviet dictator Josef Stalin. Here his picture dominates a 1940s shop window in communist-ruled Czechoslovakia.*

 Spotlight

The Terms of the Treaty
The North Atlantic Treaty of 1949, which is still in effect today, starts by stating that the member governments "desire to live in peace with all peoples and all governments." It maintains that they are determined "to safeguard the freedom, common heritage and civilization of their peoples, founded on the principles of democracy, individual liberty and the rule of law." The NATO countries agree to "promote stability and well being" and to "unite for collective defense."

The treaty also states that the original members can invite any other European country to join NATO, if appropriate.

Shared Values

All member countries stressed shared heritage and common values as the main reason for the alliance. They did not want to be seen as an aggressive military alliance alone. However, the formation of the alliance seemed like an act of aggression to the Soviet Union.

When the North Atlantic Treaty was signed, no one knew for certain how important the organization would become. The United States remained hesitant about any real military commitment to the defense of Western Europe, but this soon changed as fears of communism increased. As an organization that united the governments of several countries, NATO became a dominant force in international politics.

3. NATO in the Cold War

For forty years after its creation in 1949, NATO was a key player in the armed standoff with the Soviet Union known as the Cold War. Its military forces became the most powerful ever possessed by an alliance in the history of the world.

Soon after the NATO treaty was signed, the Soviet Union exploded its first atom bomb; China became a communist country; and war broke out in Korea between the communist North and the American-backed South. NATO became more convinced than ever of the threat posed by communism, and in response, countries strengthened their commitment. American troops and aircraft were stationed in Western Europe, and American nuclear weapons were committed to the defense of European allies.

▲ *The nuclear age. Top, the first hydrogen bomb is tested. Above, schoolchildren in Kansas practice their survival drill for use in case of nuclear attack.*

▲ British prime minister Clement Attlee meets American bomber crews stationed in East Anglia in 1949.

▲ French president Charles de Gaulle made NATO close its military bases and headquarters in France.

As the U.S. military presence in Europe increased, the alliance expanded to increase its strength. In 1952, Greece and Turkey joined, although neither country was a well-established democracy. West Germany joined the alliance in 1955, an extremely controversial move. Most of the NATO countries had fought against Germany in the two World Wars, and many people found it hard to accept the Germans as allies. The Soviet Union saw the inclusion of West Germany in NATO as a provocative act and responded by forming the Warsaw Pact, a military alliance with its fellow communist countries in Eastern Europe.

▼ CND campaigners in the 1960s protest nuclear weapons.

Nuclear Weapons

NATO and the Warsaw Pact countries became involved in a dangerous "arms race," with each side competing for more weapons of mass destruction. Rather than making the two sides feel more secure, these arms only increased the fear of war. This was especially true of nuclear weapons. By the 1960s, both the United States and the Soviet Union had developed vast nuclear armories capable of destroying most of the world's population.

NATO was committed to using nuclear weapons, if necessary, to defeat a Soviet invasion in Europe. However, many people in NATO countries protested nuclear weapons, arguing that

Spotlight

De Gaulle Against NATO

General Charles de Gaulle, president of France from 1958 to 1969, resented NATO because he saw it as a threat to France's independence. He insisted that France should have its own nuclear weapons, an "independent nuclear deterrent" under French control.

Spotlight

The NATO alliance has always been restricted to Europe and the North Atlantic — it does not commit its members to supporting one another anywhere else in the world. When the United States became involved in the Vietnam War in the 1960s, for example, other NATO countries did not send troops to help their ally. The war was in Southeast Asia, and was therefore not within NATO's territory.

they threatened the future of humanity. The British Campaign for Nuclear Disarmament (CND) was a particularly effective anti-nuclear movement. Protests were especially intense in the 1980s, when American cruise missiles with nuclear warheads were taken to Western Europe to counter the threat of Soviet warheads.

Divisions in the Alliance

There were at times sharp differences among NATO members. Greece and Turkey were generally on hostile terms with one another and came close to open warfare on several occasions. More importantly, there were tensions between European countries and the United States. Many Europeans, including France's president Charles de Gaulle, viewed NATO as an organization used by the Americans to dominate Europe. Some Americans, on the other hand, said that Europeans should contribute more money to their own defense.

In 1966, de Gaulle withdrew France from the military side of NATO. All NATO bases on French soil were closed down, and the organization was forced to relocate its headquarters from Paris to Brussels. Despite this, France remains committed to the defense of its allies, so in practice it has hardly changed its participation in the alliance.

▲ Politics forgotten, European fans were delighted when rock legend Elvis Presley served as a soldier at a NATO base in Europe.

▲ *A Royal Navy Polaris submarine, part of NATO's nuclear armory*

The End of the Cold War

Over the years the existence of two armed camps — NATO on one side and the Warsaw Pact countries on the other — became an accepted part of international politics. By the late 1960s, few people seriously expected the Soviet Union to invade Western Europe. Moves were made to reduce tensions and the risk of nuclear war. But this "detente," or relaxing of attitudes, failed to solve the fundamental problem of a divided Europe, and nuclear arsenals continued to grow. Then, in 1985, Mikhail Gorbachev became ruler of the Soviet Union.

▲ *The Berlin Wall divided East Berlin and West Berlin from 1961 to 1989.*

▸ *U.S. president Ronald Reagan and Soviet leader Mikhail Gorbachev (left) agreed to make cuts in nuclear arsenals.*

Gorbachev wanted to reform the communist system and end the Cold War. In 1987, he reached a landmark agreement with U.S. president Ronald Reagan to reduce nuclear weapons. Over the next two years, communist rule in Eastern Europe literally fell apart, culminating in the dramatic breaking down of the Berlin Wall, which had divided communist East Berlin from democratic West Berlin for 28 years. Two years later, in late 1991, communist rule ended in the Soviet Union. The large country broke up and was replaced by an independent Russia, Ukraine, and 15 other states.

NATO appeared to have fulfilled its primary objective, but it soon became clear that although the type of threats to peace had changed, the need for security and peace had not gone away.

NATO was originally formed to resist the threat of war posed by the Soviet Union. By the end of 1991, the Soviet Union no longer existed. The communist Warsaw Pact was disbanded, but **NATO** did not break up. Instead, more countries than ever were clamoring to join the organization.

Until the 1990s, NATO's focus had been on mutual defense — the part of the treaty stating that all the allies would come to the aid of any member country that was under attack. After the end of communist rule in Eastern Europe and the breakup of the Soviet Union, however, there was no enemy who could realistically attack the NATO allies.

At the same time, however, Europe became less stable. Wars broke out in Yugoslavia and the territories of the former Soviet Union. Throughout the areas once under communist rule, new governments faced severe political and economic problems.

▼ *Social and economic conditions in post-communist Russia were very poor. Below, a beggar sleeps in a Moscow street. Bottom, people pick through garbage at a dump.*

▲ NATO secretary-general Lord Robertson (right) welcomes a representative of Croatia into the Partnership for Peace.

NATO's focus shifted from mutual defense to other parts of the original treaty, which said that member countries would promote "conditions of stability and well-being," preserve "peace and security," and strengthen "free institutions" — that is, democratic government. NATO decided that its main role would be to encourage peace, security, and stability throughout Europe, especially in the large areas previously under communist rule.

Partnership for Peace

In 1994, NATO asked European countries outside the alliance to join a Partnership for Peace. Today, almost all non-NATO countries in Europe have joined the partnership. Members include Russia, Ukraine, and the other republics that had made up the Soviet Union; the other former members of the Warsaw Pact; and previously neutral countries, such as Austria, Finland, Ireland, Sweden, and Switzerland.

✔ Checklist

Partnership for Peace

In 2000, twenty-six non-NATO countries were part of the Partnership for Peace:

Albania	Kazakhstan
Armenia	Kyrghyz
Austria	Republic
Azerbaijan	Latvia
Belarus	Lithuania
Bulgaria	Moldova
Croatia	Romania
Estonia	Russia
Finland	Slovakia
Former	Slovenia
Yugoslav	Sweden
Republic of	Switzerland
Macedonia	Turkmenistan
(FYROM)	Ukraine
Georgia	Uzbekistan
Ireland	

Although the Soviet Union and its communist allies no longer exist, their armies and weapons are still there — including nuclear weapons. These armies were trained to fight against NATO and defend communism. Through the Partnership for Peace, NATO hopes to break down the barriers between itself and its former enemies, and ensure that these once communist armed forces will adapt to democracy.

The Partnership for Peace has encouraged cooperation among the armed forces of the countries involved, especially by arranging joint training exercises. The focus has been on preparing for peacekeeping missions and helping cope with natural or other disasters.

▼ An American sergeant shows an Estonian soldier how to defuse a landmine during a Partnership for Peace military exercise in 1997.

▲ *American and Russian soldiers work together in the peacekeeping Stabilization Force in Bosnia.*

Turning Enemies into Allies

As part of its new initiative for peace, NATO decided to start inviting former Warsaw Pact countries to become full members of the alliance. This step was not popular with all its members. In 1997, three countries — the Czech Republic, Hungary, and Poland — were invited to join NATO. They were selected because they had become sufficiently similar to the countries of the West — that is, they had stable democratic governments and modernizing economies. After two years of negotiations, the three countries joined NATO.

Links with Russia and Ukraine, the largest countries of the former Soviet Union, were much harder to develop. Both still had nuclear weapons and, especially in Russia, an active suspicion of NATO. The alliance made great efforts to convince the Russians of its goodwill. In 1997, a NATO-Russia Permanent Joint Council was set up to provide a formal structure for building trust and cooperation between NATO countries and Russia.

 Spotlight

Peacekeeping in Bosnia

NATO's new peacekeeping role in Europe has been demonstrated in Bosnia. A war that had broken out when Bosnia was trying to gain independence from Yugoslavia in 1992 was ended by a peace agreement in 1995. NATO sent an Implementation Force (IFOR) into Bosnia to prevent further fighting and help create stability so that peace could be achieved and democracy could begin to take root. Other countries joined in the effort, including Russia and non-NATO members of the Partnership for Peace. In 2000, many of these troops were still in Bosnia as a Stabilization Force (SFOR), committed to preserving the peace, helping displaced people return home, and seeking out war criminals.

▼ *U.S. secretary of state Madeleine Albright looks on as the Czech Republic, Hungary, and Poland sign up for NATO in 1999.*

The Mediterranean Dialog

In 1994, NATO launched the Mediterranean Dialog, which set up regular discussions between the alliance and countries in North Africa and the eastern Mediterranean, including Algeria, Egypt, Israel, Jordan, Morocco, and Tunisia. The goal was to encourage peace and stability throughout the Mediterranean region, as instability in these areas could easily spill over into the NATO countries.

Growing Pains

The expansion of NATO to include the Czech Republic, Hungary, and Poland in 1999 has raised problems for the organization. In the early 1990s, the armies in the three new member states consisted of poorly paid and demoralized troops equipped with old Soviet weapons. Few spoke English or French, the two official languages of the alliance. This situation has improved, but NATO has had to adapt to help the new country members while enjoying the benefits of the extra security their inclusion brings.

▼ *Representatives of NATO and non-NATO countries meet at the NATO headquarters in Brussels for regular talks about matters of general concern.*

Russia was wary of any eastward expansion of NATO, seeing it as a threat to its own security. To calm Russian fears, NATO agreed that no foreign troops or nuclear weapons would be stationed on the territory of the new allies. Instead, NATO would use a rapid reaction force to defend the new members if they were ever attacked.

In 1999, fifty years after it was founded, **NATO fought a war. Troops attacked Serbia from the air with missiles and bombs. The move was provoked by large-scale violation of human rights in Kosovo. NATO countries feared the troubles and unrest would spread throughout the region. The resulting conflict was highly controversial.**

▼ *President Slobodan Milosevic greets his supporters. He led Serbia into conflict with NATO over Kosovo in 1999.*

The Kosovo war stemmed from the breakup of the communist country of Yugoslavia in the early 1990s. Bosnia-Herzegovina, Croatia, Macedonia, and Slovenia declared themselves independent countries. Yugoslavia then existed as two republics, Montenegro and

Led by President Slobodan Milosevic, the Serbs opposed the independence of the former Yugoslav republics, especially Bosnia. The war in Bosnia between Croats, Muslims, and Serbs saw tens of thousands of people die, many killed in cold blood during "ethnic cleansing" operations. The United Nations originally led international efforts to stop the terrible fighting. NATO soon became involved

The Path to War

In 1998, the focus of the war in the former Yugoslavia shifted to Kosovo, a province of Serbia where the majority of the population were Albanians and the minority were Serbs. The Kosovo Liberation Army (KLA), an Albanian guerrilla movement, was fighting for independence from Serb rule. Faced with evidence of mass killings of Albanians by Serb forces, and thousands more being driven from their homes, NATO leaders threatened Serbia with air attacks. The Serbs backed down and agreed to take part in peace talks organized by NATO member states.

Difficult peace negotiations went on until March 1999. NATO repeatedly threatened air attacks to make Serbia accept the presence of a NATO-led peace force in Kosovo. When it became clear that Serbia would not accept NATO's peace initiatives, and the number of Kosovar Albanians forced from their homes had topped 250,000, NATO leaders carried out their threat. NATO air attacks began on March 24, 1999. Serbia responded by launching an offensive against the Kosovar Albanians. About half a million Albanians fled from their homes to refugee camps around the world.

▲ Former British minister of defense George Robertson favored NATO action in Kosovo.

◀ Kosovar refugees driven from their homes by Serb forces

While the massacres of Albanians continued, NATO bombs and missiles also killed about five hundred civilians, including Kosovar Albanians whom NATO was intending to help. In one incident, NATO mistakenly destroyed the Chinese Embassy in the Serb capital, Belgrade, causing a diplomatic storm.

Arriving at a Peace Settlement

NATO's involvement in the Kosovo conflict led to tension between NATO and Russia, since the Russians have traditional links with the Serbs. Negotiators finally agreed on a peace deal through an independent initiative — a joint approach to Serbia by Russia and the European Union. In June 1999, Serbia agreed to withdraw its forces from Kosovo, grant the province self-government, let refugees return to their homes, and allow a peacekeeping force to occupy Kosovo.

▶ *The Chinese Embassy in Belgrade, hit by a NATO air strike.*

▼ *British troops are welcomed as liberators in Kosovo.*

While NATO leaders described the war as a success, their task was not finished. The United Nations asked NATO to deploy a peacekeeping force, including Russian troops, to uphold the peace settlement and create stability in the area. This force, known as KFOR (Kosovo Force), was faced with huge problems. It had to try to restore security and order for both the Albanian and Serb people of the area, and help rebuild the province. The scale of this task put considerable strain on its staff and resources.

▶ *British prime minister Tony Blair saw the Kosovo war as a battle for human rights.*

▼ *A protest in Belgrade against NATO's air strikes on Serbia*

Criticism of NATO's Role in Kosovo

NATO leaders viewed the Kosovo war as being motivated largely by a desire to uphold human rights. It was their duty to intervene to protect the Kosovar Albanians from Serb violence and oppression. Said U.S. president Bill Clinton, "We cannot simply watch as hundreds of thousands of people are brutalized, murdered, and forced from their homes — all in the name of ethnic pride."

But some critics of the war saw it as an act of aggression, the bullying of a small nation (Serbia) by the major powers with their powerful weapons. NATO leaders disliked the Serbian leader Slobodan Milosevic and were glad of his political downfall in October 2000.

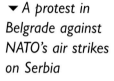

◀ *This passenger train was hit by a NATO missile, killing or injuring dozens of civilians.*

Several times during the Kosovo war, NATO aircraft struck civilian targets, including bridges, a television station, and convoys of Kosovar refugees. Criticizing NATO, Human Rights Watch declared that some of these were not appropriate military targets under international law. As a result, unnecessarily large numbers of civilians were killed — about five hundred in all.

Human Rights Watch also believed that KFOR, led by NATO, was not experienced enough in law enforcement to ensure a feeling of security for people after the war. As a result, most Serbian people left Kosovo, even though the peacekeepers were on the streets. This became another tragedy in the form of more refugees.

▲ *The headquarters of Milosevic's Socialist Party in Belgrade burns after a NATO air attack.*

Since the Cold War ended in the late 1980s, many people have questioned whether NATO is still necessary. Unfortunately, conflicts still break out in Europe and elsewhere. They just tend to be more scattered.

At the beginning of the twenty-first century, NATO forces were training for, and carrying out, a range of new roles, from peacekeeping to organizing relief efforts in response to natural disasters. These roles require smaller, mobile forces rather than the military might of the past.

▼ British troops serve with the Stabilization Force in Bosnia. This kind of peacekeeping operation is one of the main justifications for the continued existence of NATO.

For many NATO leaders, the Kosovo war pointed to a new mission for NATO as the upholder of human rights in Europe. Not all NATO states were supportive of further military intervention in trouble spots, however. Some argued the use of force caused more problems than it solved, increasing instability and making a permanent peace harder to establish.

▼ *Bosnian, Croatian, and Serbian leaders met to end the Bosnian conflict in 1995. But the situation in the Yugoslav region has remained unsettled.*

The EU and NATO

Because of its military might, the United States has supplied much of the weaponry and armed forces when NATO has intervened in a crisis. However, the European NATO members' dependence on the United States is an ongoing issue, as it has been since the founding of the organization. As ties between European Union (EU) countries — most of whom are also NATO members — became closer during the late 1990s, discussions centered around the creation of a European Rapid Reaction Force. This would be a multinational force that could handle crises without the aid of the United States.

However, most EU member states have very little in the way of military resources compared with the United States. The proposed force, therefore, would not be able to form a "Euro-Army" as an alternative to NATO. Instead, it would use NATO resources for its own operations in cases where the United States did not want to get involved.

Old Divisions Die Hard

NATO's goal of spreading security and stability across Europe was put at risk with the expansion of the alliance in the 1990s. Fears that the expansion would harm relations with Russia have so far proved unfounded.

However, the Russians are aware of their position as NATO's only potential major enemy. Anything that increases NATO's power makes the Russians feel less, rather than more, secure. The Baltic republics of Estonia, Latvia, and Lithuania are eager to join NATO, but the Russian government opposes any such move. Critics of NATO argue that it threatens to once again divide Europe into two hostile camps, this time with the "Iron Curtain" along the eastern borders of the Czech Republic, Hungary, and Poland.

▶ *Vladimir Putin, who became president of Russia in 2000, might reassert his country's military power.*

The Chechnya Crisis

In the winter of 1999–2000, Russia crushed an independence movement in the republic of Chechnya, a part of the Russian Federation. Russia's conduct was widely criticized, with well-substantiated allegations of massacres and other atrocities committed by Russian forces. Unlike in Kosovo, NATO did not intervene to stop the massacres, nor did it put pressure on Russia to change its behavior.

The events in Chechnya (see panel) challenged NATO's commitment to uphold human rights. Russia is too powerful to be threatened without the risk of a major war. NATO leaders reluctantly accepted that they had to work with, rather than against, Russia.

▶ *Russian soldiers in action in Chechnya*

▼ *Russia used tanks against lightly armed Chechen fighters.*

The Future of NATO

NATO maintains substantial nuclear and conventional forces, although defense spending is high in only a few member countries. However, to act independently, every country in NATO would have to increase its defense spending. Today, NATO's work is based on the view that peace is best maintained by creating forums for discussion and negotiation and through cooperation among nations. The future of the alliance will depend on the continuing belief of the people of Europe and North America — and their leaders — that NATO is essential to guarantee international security.

● Spotlight

Disaster Relief

NATO is increasingly involved in disaster relief. In 1999, NATO opened a Euro-Atlantic Disaster Relief Coordination Center to head up emergency and relief operations in the event of a disaster. Its forces can distribute food or other aid, for example, to victims of flooding in Ukraine in 1999 or to refugees fleeing the fighting in Kosovo in 1999.

▲ Victims of the Ukraine floods in 1999. NATO troops helped in the relief of this natural disaster.

▼ The fall of the Berlin Wall in 1989: a victory for democracy — and for NATO?

● Spotlight

In 1997, proposing that his country join NATO, Polish foreign minister Bronislav Geremek made a powerful case for the continued existence of the alliance:

"We would prefer to live in a Europe with no arms and no alliances. But we do live in a world where military power remains the ultimate guarantor of security. NATO is an alliance which has managed to put its immense military might in service of fundamental values and principles that we share. NATO can make Europe safe for democracy. No other organization can replace the alliance in this role."

arms race	the situation that arises when two or more countries or alliances try to produce more powerful arms than the other, leading to the development of even more powerful weapons in even larger numbers	**dictatorship**	rule by an individual who has absolute power over his or her country
atrocities	extremely cruel, brutal acts, especially against defenseless people	**ethnic cleansing**	mass killing, or driving all people of a particular ethnic group out of the place where they live, by force or terror
Cold War	the armed confrontation between the United States and its allies on one side and the Soviet Union and its allies on the other, which lasted from the late 1940s to the late 1980s	**guerrillas**	lightly armed fighters usually engaged in a war against the government of their own country
		heritage	traditions and values handed down from the past
communism	a political and economic system whereby everyone works for the common good rather than individual gain. As the government is not elected democratically, it is almost impossible for the people to remove bad government.	**human rights**	freedoms that all humans have the right to enjoy, wherever they live and whatever their government
		Iron Curtain	a phrase coined by British statesman Winston Churchill to describe the fortified line that divided Western Europe from communist-ruled Eastern Europe from the late 1940s to the 1980s
democracy	political system in which the people elect their rulers	**military alliance**	an agreement between countries to fight together against a common enemy
detente	reduction of tensions that might lead to war		

neutral	not taking sides in a conflict or confrontation
nuclear weapons	weapons of immense destructive power using the energy from the nuclei in atoms
referendum	a vote by a country or state's population on a specific issue
self-government	the right of the people of a particular area within a country to choose their own government but not have full independence from that country
Soviet Union	the communist nation that until 1991 ruled a vast area of Europe and Asia, including Russia, Ukraine, Belarus, Georgia, Armenia, and Kazakhstan
terrorist	a member of a political movement that uses violence to achieve its goals
Warsaw Pact	military alliance set up in 1955 by the Soviet Union and the communist countries of Eastern Europe

NATO Headquarters
Boulevard Leopold III
1110 Brussels
Belgium

NATO Web site:
www.nato.int
This site provides a brief history of the organization, regularly updated information on current topics such as Kosovo, details of NATO's policy agreements, and access to an extensive archive of documentary material.

Human Rights Watch
1630 Connecticut Avenue NW
Suite 500
Washington, DC 20009

Human Rights Watch Web site:
www.hrw.org
This site provides a large body of research on human rights issues around the world.

Amnesty International
322 8th Avenue
New York, NY 10001

Amnesty International Web site:
www.amnesty.org
This site deals with human rights issues worldwide.

Albright, Madeleine 18
Amnesty International 31
arms race 12, 30
atom bomb 11
Attlee, Clement 12

Belgium 4, 6, 7, 10
Berlin 8, 9, 14
Berlin Wall 14, 29
Bevin, Ernest 9
Blair, Tony 23, 26
Bosnia and Herzogevina 5, 18,
  20
Brussels 6, 13, 19, 31
Brussels Treaty 10

Campaign for Nuclear
  Disarmament (CND) 12–13
Canada 4, 10
Chechnya 28
Churchill, Winston 30
Clinton, Bill 7
Cold War 5, 11, 14, 25, 30
communism 5, 8–10, 11, 12, 14,
  15, 17, 30
cruise missiles 13
Czechoslovakia 8, 10
Czech Republic 4, 18, 19, 27

de Gaulle, Charles 12, 13
democracy 5, 6, 8, 9, 12, 14, 17,
  30
Denmark 4, 10

ethnic cleansing 20, 30
European Union (EU) 26

France 4, 9, 10, 13
Franco, Francisco 6

Geremek, Bronislav 29
Germany, East 8
Germany, West 4, 8, 12
Gorbachev, Mikhail 14

Greece 4, 6, 12, 13

headquarters, 6
Hungary 4, 8, 18, 19, 27
hydrogen bomb 11

Iceland, 4, 10
Implementation Force (IFOR)
  18
Iron Curtain 8, 27, 30
Italy 4, 10

Korea 11
Kosovo 5, 20–24, 28, 29
Kosovo Force (KFOR) 22

logo/symbol, NATO 5
Luxembourg 4, 10

Marshall Plan 9
Mediterranean Dialog 19
members, list of 4
military committee 7
Milosevic, Slobodan 20, 23, 24

NATO-Russia Permanent Joint
  Council 18
Netherlands 4, 10
North Atlantic Council (NAC)
  6
North Atlantic Treaty 8, 10
Norway 4, 10
nuclear weapons 11, 12, 13, 14,
  18, 19, 28, 31

Partnership for Peace 16–18
Poland 4, 8, 18, 19, 27, 29
Portugal 4, 10
Presley, Elvis 13
Putin, Vladimir 27

Reagan, Ronald 14
Robertson, Lord George, 6, 16,
  21

Russia, 4, 5, 14, 15, 16, 18–19,
  22, 27, 28

secretary-general 6, 16
Serbia 5, 20–21, 23
Soviet Union 4, 5, 8–9, 11–12,
  14, 15, 16, 18, 30, 31
Spain 4, 6
Stabilization Force (SFOR) 18,
  25
Stalin, Josef 10

Truman, Harry S. 8, 9
Turkey 4, 6, 12, 13

Ukraine 4, 5, 14, 16, 18, 29
United Kingdom 4, 9, 10
United States 4, 5, 7, 8, 9, 10, 11,
  12, 13, 18, 26

Vietnam War 13

Warsaw Pact 5, 12, 14, 15, 18,
  31
Western Union 10
World War II 8, 12

Yugoslavia 15, 18, 20, 21